Pocket
Earth
Wisdom

Pocket Earth Wisdom

Wise words to inspire you to sit up, listen and take action to save our planet

Hardie Grant

BOOKS

CONTENTS

"Who can stop climate change? We can. You and you and you, and me."

DESMOND TUTU

Our planet is in danger. Sea levels are rising, natural habitats are being destroyed and the global temperature is rising. It cannot be denied any more: climate change is real, and it isn't going away anytime soon. There's no planet B, people.

There is some good news though: we can halt climate change if we work together. Let the words of activists, world leaders, environmentalists and even your favourite film stars, of all ages and backgrounds, inspire you and those around you. A call of arms so each and every one of us can make changes in our lives, however big or small, to ensure future generations can enjoy our planet as much as we have. The power of the collective cannot be denied and, if we unite together, we can change our future and that of the generations to come. It impacts each and every one of us.

Now more than ever it's imperative that we take a stand and speak out. The time is now.

CLIMATE

CHANGE

"Climate change is not hysteria - it's a fact."

LEONARDO DICAPRIO

"In the 20 years since I first started talking about the impact of climate change on our world, conditions have changed far faster than I ever imagined."

SIR DAVID ATTENBOROUGH

"Climate change is not just a problem for the future. It is impacting us every day, everywhere."

VANDANA SHIVA

"Next time you meet a climate-change denier, tell them to take a trip to Venus; I will pay the fare."

STEPHEN HAWKING

"Climate change
knows no borders."

ANGELA MERKEL

"Climate crisis is the era that we inhabit. It forms the backdrop to our lives, a constant soundtrack which will only get louder."

ELLIE GOULDING

"I don't believe that there's anybody in this world that can deny science."

PRINCE HARRY

"Climate change is real. That is not fake news. And we cannot, cannot be the generation the history of the world will look back on and wonder why they didn't do everything humanly possible to solve the biggest crisis in our time."

EDDIE VEDDER

"This is not some sort of doomsday argument, we're destroying the planet."

CHRIS HEMSWORTH

"Climate change is one of the most defining issues of our time, one that threatens our very existence on Earth."

PHARRELL WILLIAMS

"The planet is dying. We have to be warriors."

MARINA ABRAMOVIĆ

"Climate change is in everybody's backyard."

ROBERT REDFORD

PLANET

EARTH

"Every time I make a decision, I think, 'How is this going to affect the environment?'"

ALICIA SILVERSTONE

"The Earth is for all beings, today and tomorrow."

VANDANA SHIVA

"As people alive today, we must consider future generations: a clean environment is a human right like any other."

HIS HOLINESS THE 14TH DALAI LAMA

"One thing leads to the other. Deforestation leads to climate change, which leads to ecosystem losses, which negatively impacts our livelihoods - it's a vicious cycle."

GISELE BÜNDCHEN

"If we save our wild places, we will ultimately save ourselves."

STEVE IRWIN

"How can our brains be so different? How can you not notice all of the things changing?"

CHER ON CLIMATE CHANGE DENIERS

"Surely, it is our
responsibility to do
everything within our
power to create a planet
that provides a home not
just for us, but for all life
on Earth."

SIR DAVID ATTENBOROUGH

"We have to understand our environment so [our] impact on this earth can really make a difference."

JADEN SMITH

"When one country faces a climate disaster, we all face a climate disaster."

CATE BLANCHETT

"We say we protect the things we love and I love the ocean and I love nature, so it's my turn."

JACK JOHNSON

"I think the environmental movement can only be successful if we are simple and clear and make it a human story."

ARNOLD SCHWARZENEGGER

"It is time to stand up and save our home."

EMMA THOMPSON

POLITICAL

CLIMATE

"We cannot condemn our children, and their children, to a future that is beyond their capacity to repair."

BARACK OBAMA

I urge African Leaders and world leaders to take into consideration that all of us will be affected by climate change. No one should be left behind.

VANESSA NAKATE

"This is not an elitist issue, this is a quality of life issue."

ALEXANDRIA OCASIO-CORTEZ

"Climate change has never received the crisis treatment from our leaders, despite the fact that it carries the risk of destroying lives on a vastly greater scale than collapsed banks or collapsed buildings."

NAOMI KLEIN

"If enough of us realise we think the same, and want to change the world for the better, we could form an opposition strong enough to sway the government now, couldn't we?"

VIVIENNE WESTWOOD

"It's not about politics, it's about our moral obligation to one another, to our children and to the future generations who will one day inherit this earth."

SIGOURNEY WEAVER

"The Earth is definitely haemorrhaging: the oil, the burning of the forests. It's upon us and world leaders don't seem to want to try and come together."

PIERCE BROSNAN

"Climate change is not a coincidence or a scientific anomaly. Climate change is a consequence. It is a consequence of our unsustainable way of life."

ALEXANDRIA OCASIO-CORTEZ

"The question for all of us is what side of history will we choose to sit on."

JACINDA ARDEN

"I've starred in a lot of science fiction movies and, let me tell you something, climate change is not science fiction, this is a battle in the real world, it is impacting us right now."

ARNOLD SCHWARZENEGGER

TIME

TO ACT

"What you do makes a difference, and you have to decide what kind of difference you want to make."

JANE GOODALL

"We need to solve the climate crisis, it's not a political issue, it's a moral issue. We have everything we need to get started, with the possible exception of the will to act, that's a renewable resource, let's renew it."

AL GORE

"This is the time for everyone because this isn't one war or one issue. This is everything. It impacts every aspect of our lives, our health, our economy, our jobs, how we live, how we can move around, whether we're going to be forced to move."

JANE FONDA

"Our house is still on fire.
Your inaction is fuelling
the flames by the hour."

GRETA THUNBERG

"The time to answer the greatest challenge of our existence on this planet... is now."

LEONARDO DICAPRIO

"It will be the fight of the century, of our capacity to invent new ways to live and do, sustainably."

EMMANUEL MACRON

"We have allowed ourselves to become totally dependent, and are guilty of ignoring the warning signs of pending disaster. It is time to act."

DESMOND TUTU

"I do have hope but hope without action isn't worth much."

DON CHEADLE

"Right here, right now is where we draw the line. The world is waking up. And change is coming."

GRETA THUNBERG

"We're perfectly capable of sorting this out, we just need to get it done. If we don't do that, we get what we bloody well deserve."

EMMA THOMPSON

"Together, we must protect
and restore nature, clean
our air, revive our oceans,
build a waste-free world
and fix our climate."

PRINCE WILLIAM

"It is our failure to some
extent that we have not
managed to mobilise
7 billion people against
the few hundred who are
making this happen."

GEORGE MONBIOT

"There's no shortage of issues that one can get involved in that impacts climate, because everything impacts climate."

AMERICA FERRERA

"We still retain the ability to avoid truly catastrophic, civilisation-ending consequences if we act quickly."

AL GORE

HUMAN

KIND

"We all know what to do.
Why don't we do it?"

WANGARI MAATHAI

"We're about to die if we don't change."

BILLIE EILISH

"People need to wake the f*ck up."

JAMES CAMERON

"It is part of our responsibility towards others to ensure that the world we pass on is as healthy, if not healthier, than when we found it."

HIS HOLINESS THE 14TH DALAI LAMA

"Our children and grandchildren should not have to pay the cost of our generation's irresponsibility."

POPE FRANCIS

"Do we want to go down in history as the people who did nothing to bring the world back from the brink in time to restore the balance when we could have done? I don't want to."

PRINCE CHARLES

"The scale of the challenge
requires a global effort,
but our individual actions
add up, and matter."

DON CHEADLE

"Who can stop climate change? We can. You and you and you, and me."

DESMOND TUTU

"The question is not whether climate change is happening, but whether, in the face of this emergency, we ourselves can change fast enough."

KOFI ANNAN

"Perhaps in a few hundred years, we will have established human colonies amid the stars, but right now we only have one planet, and we need to work together to protect it."

STEPHEN HAWKING

"The real trigger in people's minds should be that this is going to cost us a lot more later than it is now."

JAMES CAMERON

"This movement is not one person, or one group. This movement is all of us and we need to make sure we value those who are disproportionately affected by this crisis. And allow those who are affected to lead."

ISRA HIRSI

LOOKING

FORWARD

"You should never give up hope."

THOM YORKE

"How we behave is going to have a serious impact on the world that these next generations are going to inherit and I think it's incumbent on all of us to do our part to try to ensure that's a world that's as good as the one we enjoy."

MATT DAMON

"I'm optimistic about climate change because of innovation."

BILL GATES

"You have the courage to imagine a future that puts our beautiful planet first and protects every living species equally."

ELLIE GOULDING

"It's this: Don't try to live up to a super high standard, but be aware. Be aware of how you can contribute. That's how we'll realistically get it done."

PHARRELL WILLIAMS

"I'm hopeful, because I think the tide is turning, and many - particularly young people - understand the existential threat we face and are prepared to do something about it."

MARK RUFFALO

"We have to remain optimistic. It can't be all about problems - it has to be about solutions."

IAN SOMERHALDER

"Addressing the climate challenge presents a golden opportunity to promote prosperity, security and a brighter future for all."

BAN-KI MOON

"I do believe that if every person does something small, you can make a big impact."

JESSICA ALBA

"To survive, it's going to take everything we've got. And everyone we know."

SAM KNIGHTS

1 Million Women. 2015. '5 Celebrities Who Love Sustainable Fashion'. 1 Million Women [online] www.1millionwomen.com.au

Annan, K. 2006. 'Climate change is not just an environmental issue'. Independent [online www.independent.co.uk

Baynes, C. 2019. 'Ocasio-Cortez delivers devastating address to congress after Republican calls Green New Deal elitist: "People are dying"'. Independent [online] www.independent.co.uk

BBC News, 2019. 'Prince Harry 'troubled' by climate change deniers'. BBC News [online] www.bbc.co.uk

BBC News. 2019. 'Ellie Goulding on climate change: "The backlash grows ever uglier. BBC News [online] www.bbc.co.uk

BBC News. 2020. 'Greta Thunberg seeks Africa climate change action'. BBC News [online] www.bbc.co.uk

Bedigan, M; Parnaby, L. 2019. 'Ellie Goulding says fans need climate change answers'. The Ecologist [online] www.theecologist.org

Bennet, J. 2015. 'We Need an Energy Miracle'. The Atlantic [online] www.theatlantic.com

Brown, A. 2019. 'Billie Eilish isn't stressing over the Grammys. She's busy worrying about the end of the world'. Los Angeles Times [online] www.latimes.com

Buchanan, B. 2017. 'Eddie Vedder Reveals Why Climate Change Is 'Biggest Crisis Of Our Time'. Alternative Nation [online] www.alternativenation.net

Carrara, J. 2019. 'Dr Jane Goddall's Best Quotes'. Eco-Age [online] www.eco-age.com

Carrington, D. 2015. 'Pharrell Williams calls on leaders to deliver green jobs at the UN climate summit'. The Guardian [online] www.theguardian.co.uk

Cheadle, D. 2015. 'Don Cheadle: Protect the World's Vulnerable From Climate Change'. Time [online] www.time.com

Chin, K; Shamo. L; Abadi; M. 2020. 'Jane Fonda explains why she's willing to get arrested in the fight against climate change'. Business Insider [online] www.businessinsider.com

Christensen, M. 2017. 'Chris Hemsworth on being Thor and living in LA: "It's suffocating and you stop becoming a person"'. GQ [online] www.gq.com.au

BBC News. 2019. 'David Attenborough climate change TV show a "call to arms"' BBC News [online] www.bbcnews.co.uk

Davies, P. 2019. '30 of the Most Impactful Climate Change Quotes'. Curious Earth [online] curious.earth

Davis, A. 2016. 'America Ferrera On One Reason Living Under Trump Is "Terrifying"'. Refinery29 [online] www.refinery29.com

DiCaprio, L. 2014. 'Leonardo DiCaprio at the UN: "Climate change is not hysteria – it's a fact"'. The Guardian [online] www.theguardian.co.uk [USED TWICE]

Donnelly, M. 2019. 'James Cameron on the Climate Crisis: "People Need to Wake the F-k Up"'. Variety [online] www.variety.com

Doyle, J. 2014. 'The Governator's got a new foe – Climate Change'. The Globe and Mail [online] www.theglobeandmail.com

Extinction Rebellion. 2019. This is Not A Drill. London: Penguin Random House

Gore, A. 2019. 'Al Gore: The Climate Crisis Is the Battle of Our Time, and We Can Win'. The New York Times [online] www.nytimes.com

Groves, N. 2015. 'Marina Abramović: 'The planet is dying. We have to be warriors.'. The Guardian [online] www.theguardian.com

Hannam, P. 2014. 'Climate change "won't stop at the Pacific Islands" Angela Merkel tells Australia'. The Sydney Morning Herald [online] www.smh.com.au

Harvey, F; Ambrose, J. 2019. 'Pope Francis declares 'climate emergency' and urges action'. The Guardian [online] www.theguardian.com

Hawking, S. 2016. 'This is the most dangerous time for our planet'. The Guardian [online] www.theguardian.com

Hertzberg, R. 2018. 'Musician Jack Johnson Wages War on Ocean Plastic'. National Geographic [online] www.nationalgeographic.co.uk

Herzog, K. 2016. 'Climate change is scarier than vampires – just ask Ian Somerhalder'. Grist [online] www.grist.org

Hirsi, I. 2019, 'The climate movement needs more people like me'. Grist [online] www.grist.org

His Holiness the 14th Dalai Lama. 'A Clean Environment Is a Human Right' [online] [Accessed: 8 July 2020] www.dalailama.com

In This Climate (Trailer). 2016 [TV] Pablo Ganguli, Tomas Auksas. dir. USA: Liberatum

Irwin, Steve and Terri. 2002. The Crocodile Hunter. London: Dutton Books

Jackson, E. 2019. 'Pierce Brosnan on James Bond, his childhood in Ireland and his environmental activism'. France24 [online] www.france24.com

Klein, N. 2014. This Changes Everything: Capitalism vs. the Climate. London: Penguin Press

McDiarmid, J. 2015. 'Emma Thompson's presence adds weight to climate change march in Ottawa'. The Star [online] www.thestar.com

Nagourney, A. 2007. 'Gore Wins Hollywood in a Landslide'. The Caucus [online], www.thecaucus. blogs.nytimes.com

NBC News. 2014. 'Q&A: 'Avatar' Director James Cameron Talks Climate Change'. NBC News [online] www.nbcnews.com

NPR Staff. 2019. 'Transcript: Greta Thunberg's Speech At The U.N. Climate Action Summit'. NPR [online] www.npr.org

Orange, R. 2019. 'Ocasio-Cortez tells world's mayors drastic action needed on climate crisis'. The Guardian [online] www.theguardian.com

Oremiatzki, Y. 2015. 'Thom Yorke and George Monbiot: "We have to prepare for the inevitable failure of COP21"'. Télérama [online] www.telerama.fr

Planet Earth II. Episode 6, 2016 [TV] Vanessa Berlowitz, Mike Gunton, James Brickell, Tom Hugh-Jones. dir. United Kingdom: BBC Natural History Unit, BBC Studios

Pooley, E. 2016. 'Behind the scenes with Don Cheadle: "Climate change is real and we must act."'. Environmental Defense Fund [online] www.edf.org

Press Association. 2020. 'Russell Crowe and Cate Blanchett use Golden Globes to highlight Australia's wildfires tragedy'. the Telegraph [online] www.telegraph.co.uk

Princes Charles. 2020. 'We need revolutionary action to save the planet: full transcript of Prince Charles' Davos speech'. The Sydney Morning Herald [online] www.smh.com.au

Prince William – TED Countdown. 2020. 'This decade calls for Earthshots to repair our planet'. TED [online] www.ted.com

Redford, R. 2015. 'Interview: "Climate change is in everybody's backyard". Office of the Secretary-General's Envoy on Youth [online] [Accessed: 13 July 2020] www.un.org

Remsen, N. 2017. 'Pharrell on Living Sustainably, and Why the Kids Will Save Us All'. Vogue [online] www.vogue.com

Ruffalo, M. 2015. 'Mark Ruffalo talks climate change and what you can do about it'. Grist [online] www.grist.org

Sengupta, S. 2020. 'Greta Thunberg's Message at Davos Forum: 'Our House Is Still on Fire'. The New York Times [online] www.nytimes.co

Shiva, V. 2015. Soil Not Oil: Environmental Justice in an Age of Climate Crisis, 2015, Berkeley: North Atlantic Books

Smith, J. 2019. 'Jaden Smith, Greta Thunberg Speak at Climate Rally'. ET Canada [online] www.youtube.com

Stephen Hawking's Favourite Places. Episode 2, 2016 [TV] Ed Watkins. dir. US: CuriosityStream, Bigger Bang.

The Associate Press. 2020. 'Macron vows "fight of the century" against climate change'. Business Insider [online] www.businessinsider.com

The Big Issue, 2019. 'Vivienne Westwood: We must disrupt corrupt systems if we want to make change'. The Big Issue [online] www.thebigissue.com

The Hollywood Reporter. 2016. 'Sigourney Weaver Talks Climate Change at the Democratic National Convention'. The Hollywood Reporter [online] www.hollywoodreporter.com

Tidman, Z. 2019. 'New Zealand passes 'zero carbon' law in fight against climate change'. Independent [online] www.independent.co.uk

Tutu, D. 2014. 'Desmond Tutu: We fought apartheid. Now climate change is our global enemy'. The Guardian [online] www.theguardian.com

United Nations. 2014. 'Secretary-General's remarks at Climate Leaders Summit'. United Nations Secretary General [online] www.un.org/sg

Vidal, J. 2009. 'We know what to do. Why don't we do it?'. The Guardian [online] www.theguardian.com

Vidal, J. 2015. 'Arnold Schwarzenegger: climate change is not science fiction'. The Guardian [online] www.theguardian.com

Years of Living Dangerously. 2014 [TV] Daniel Abbasi, Joel Bach, James Cameron, David Gelber, Arnold Schwarzenegger, Jerry Weintraub. dir. USA: Showtime [USED TWICE]

Yorke, T. 2008. 'Thom Yorke: why I'm a climate optimist'. The Guardian [online] www.theguardian.com

XR Video. 2019. It's time to stand up and save our home: Emma Thompson'. Extinction Rebellion [online] rebellion.earth

Published in 2020 by Hardie Grant Books,
an imprint of Hardie Grant Publishing

Hardie Grant Books (London)
5th & 6th Floors
52–54 Southwark Street
London SE1 1UN

Hardie Grant Books (Melbourne)
Building 1, 658 Church Street
Richmond, Victoria 3121

hardiegrantbooks.com

British Library Cataloguing-in-Publication Data. A catalogue
record for this book is available from the British Library.

Pocket Earth Wisdom
ISBN: 978-1-78488-4260

10 9 8 7 6 5 4 3 2 1

Publishing Director: Kajal Mistry
Assistant Editor: Alexandra Lidgerwood
Design and Art Direction: Hannah Valentine
Illustrations: Hannah Valentine
Production Controller: Sinead Hering

Colour reproduction by p2d
Printed and bound in China by Leo Paper Products Ltd